IN THE NATIONAL INTEREST

General Sir John Monash once exhorted a graduating class to 'equip yourself for life, not solely for your own benefit but for the benefit of the whole community'. At the university established in his name, we repeat this statement to our own graduating classes, to acknowledge how important it is that common or public good flows from education.

Universities spread and build on the knowledge they acquire through scholarship in many ways, well beyond the transmission of this learning through education. It is a necessary part of a university's role to debate its findings, not only with other researchers and scholars, but also with the broader community in which it resides.

Publishing for the benefit of society is an important part of a university's commitment to free intellectual inquiry. A university provides civil space for such inquiry by its scholars, as well as for investigations by public intellectuals and expert practitioners.

This series, In the National Interest, embodies Monash University's mission to extend knowledge and encourage informed debate about matters of great significance to Australia's future.

Professor Sharon Pickering
President and Vice-Chancellor,
Monash University

QUENTIN GRAFTON

RETELLING AUSTRALIA'S WATER STORY

Retelling Australia's Water Story
© Copyright 2024 Quentin Grafton
All rights reserved. Apart from any uses permitted by Australia's *Copyright Act 1968*, no part of this book may be reproduced by any process without prior written permission from the copyright owners. Inquiries should be directed to the publisher.

Monash University Publishing
Matheson Library Annexe
40 Exhibition Walk
Monash University
Clayton, Victoria 3800, Australia
https://publishing.monash.edu

Monash University Publishing brings to the world publications which advance the best traditions of humane and enlightened thought.

ISBN: 9781922979902 (paperback)
ISBN: 9781922979926 (ebook)

Series: In the National Interest
Editor: Greg Bain
Project manager & copyeditor: Paul Smitz
Designer: Peter Long
Typesetter: Cannon Typesetting
Proofreader: Gillian Armitage
Printed in Australia by Ligare Book Printers

A catalogue record for this book is available from the National Library of Australia.

The paper this book is printed on is in accordance with the standards of the Forest Stewardship Council®. The FSC® promotes environmentally responsible, socially beneficial and economically viable management of the world's forests.

RETELLING AUSTRALIA'S WATER STORY

Most Australians know how important is fresh water. Should those living in the biggest cities forget its importance, they are reminded of it when they face water restrictions in the summer, flooded roads when a major cyclone hits, or 'No Swimming' advisories after a storm because of polluted stormwater run-off. Water, however, is much more important than many of us fully realise. Water has shaped our history and helped make Australia what it is today. Without understanding the crucial part that water has played in our nation's story, we cannot build and sustain a better Australia.

The orthodox Australian water story includes 'nation-building' water projects that helped to develop Australian industry, such as the Snowy Mountains hydro-electricity scheme; a highly water-use-efficient irrigation industry that 'saves' water; the world's most diverse and largest formal water market; and that we all enjoy access to reliable, good-quality water. Our standard water history also includes made-in-Australia policy innovations, such as government purchases of water rights from willing sellers to increase streamflows and to do 'big' basin planning—innovations that have led some to claim that Australia has the world's best practice when it comes to managing its water.

This book tells a different story. It's time to look more deeply and see what's hidden from full sight. It's time to go back into our past to foretell a better water future. It's time to retell Australia's water story.

Australia can indeed lay claim to so-called nation-building water projects. Some of these projects, however, resulted in very large water diversions that have greatly reduced the streamflow of some of our most iconic rivers.

Australia does have a water-use-efficient irrigation industry. But this has, in part, been achieved by billions of dollars in government subsidies—on average, over half a million dollars per irrigator, with some receiving millions of dollars. Importantly, and paradoxically, subsidies to increase water-use efficiency may have reduced streamflows in some of our largest rivers.[1]

And yes, water markets in the Murray–Darling Basin are world-leading in terms of value and the amount of trade. But they are unjust because virtually no water rights were allocated by the states to First Peoples.

The allocation of water licences, which then became water entitlements, also represented a multibillion-dollar transfer of public wealth to just a few thousand irrigators. While Australians in our biggest cities and towns are—justifiably—thankful for the good-quality and reliable water they enjoy, hundreds of thousands of Australians living in rural and remote locations still regularly face multiple threats. These threats include temporary boil advisories to remove microbiological pollution such as *E. coli* bacteria, and non-microbial contaminants that cause ill-health and cannot be removed by boiling, such as nitrates. Smaller communities also face the real risk of their supplies running out during an extended drought.

The stories in this book are supported by evidence. Some are confronting, but they need to be told. A few are cautionary, telling us what not to do. It is necessary to learn from all of them if we are to live sustainably and prosper in this 'Great Southern Land'.

Retelling Australia's Water Story is about rethinking what we know. It's a story that ultimately challenges our myths about water; for example, that we can 'drought-proof' Australia. It's a story that helps us to recognise, if we do not already, that water is not simply a 'resource' that only has worth when it is diverted, used and consumed, nor that it is 'wasted' if rivers flow out to sea. Water sustains us: it protects us and gives us life. For our own sake, and for our communities, and for our non-human cousins, we must also nurture and sustain it.

When you take this water journey with me, you will see the ancient water story of Australia and its First Peoples, and along the way learn practical, place-based insights into what's wrong with the 'world's best water practice' label. You will see the alternative vision of 'living waters'. You will read about the falsehood of aqua nullius,[2] that water was owned by no-one, and how water was acquired by state governments and then mostly given away to establish and support an irrigation industry. You will encounter the great diversions of water that fundamentally changed the Australian waterscape and landscape, causing great degradation. And you will also look ahead to what is coming in terms of climate change, the 'future imperfect', before contemplating a vision of how we can reconcile ourselves with our water, and it with us, to deliver water for all, justly, so that no-one is left behind—not today nor for the many tomorrows to come.

TOO LITTLE, TOO DIRTY

The haves and the have-nots of water access and decision-making have been part of Australia's water story from the arrival of the First Fleet in 1788. The newcomers wanted to make their home in Gadigal Country, in what today is known as Sydney. Fresh water was important from the very beginning. Sydney's original location at Bennelong Point in Port Jackson (site of the Sydney Opera House) was chosen because of a lack of accessible fresh water in Kamay—what today is known as Botany Bay.

The first, and initially only, source of fresh water for the new colony was known to colonists as the 'streamlet', which was fed by springs and a wetland bounded by what are today Market, Pitt, Park and Elizabeth streets. Recognising the importance of a reliable and potable source of water for the colony, governor Phillip decreed a green space of several metres on either side of the streamlet where the felling of trees, or the grazing or stocking of animals, was prohibited.[3] As the Sydney colony grew, houses on the eastern side of the streamlet, what is now Pitt Street and Macquarie Street, became the residences of the haves, including the Governor's House. Those on the western side, where the working class and the convicts resided, what is now known as The Rocks, became the home of the have-nots.

In 1790, with the colony facing a drought, governor Phillip ordered three tanks to be dug in the sandstone next to the streamlet to capture and store the limited fresh water. As a result of this construction, the streamlet became known as the Tank Stream. These water tanks were colonial Australia's first attempt at overcoming water insecurity.

In the many years since, Australians have adopted a similar approach in storing water for the 'dry', in a country with a highly variable climate—of droughts and flooding rains. A key goal has been to capture and store water when it rains to make it more available in drier periods. Today, there are some 1.7 million farm dams across Australia that are complemented by dozens of much larger dams. Some of the largest include the Hume Dam, built primarily to provide greater and more reliable streamflows for irrigation in the summer months along the Murray River; the Warragamba Dam, Sydney's largest water storage; and the Gordon Dam in Tasmania, built for hydro-electricity generation.

Governor Phillip's departure from Sydney in 1792 allowed corrupt army officers, the so-called 'Rum Corps' (because they used hard liquor—the supply of which they controlled—as the de facto currency), to become the key decision-makers in the colony. The Rum Corps overturned the green-space rules and allowed houses and pigsties to be built alongside the Tank Stream. This was the beginning of the end for the stream as a potable source of water, despite subsequent attempts to protect it from pollution. By the 1820s, the Tank Stream had become an open sewer and a major source of water-borne diseases. The stream is now buried under Sydney's

CBD, relegated to a stormwater drain that flows into Sydney Harbour.

What happened to the Tank Stream was a precursor of what was to happen to Australia's surface water, again and again, in the centuries that followed—the despoilment and diversion of streams and rivers, and the dispossession of water from First Peoples.

Recognising the importance of water, Australia's colonial governments—Victoria in 1886, New South Wales in 1896 and South Australia in 1919—vested the rights to use and control water in the Crown or the state. In the intervening period, only a tiny fraction of this water has been returned to First Peoples. By contrast, over the past forty years, most of the accessible surface water has been allocated to irrigators in the form of tradeable water rights known as water entitlements.

And should you think that a lack of justice and unsustainable water use lie in the colonial past, think again. Right here and right now, precious water sources are still being degraded to benefit the few at the expense of the many. Water continues to flow uphill to those with money, power and influence.

LIVING WATERS

As became the norm in Australia's history of colonisation, the Country of the Gadigal people was stolen without treaty. This was done under the unprincipled and patently false premise that the land and the water on which First Peoples had lived for millennia belonged to no-one (terra nullius and aqua nullius, respectively). Importantly, telling Australia's water story only from 1788 misses at least 65 000 years of continuous habitation by Australia's First Peoples, from the coasts to the deserts, from the plains to the mountains. Everywhere in the ancient landscape was settled, and every place was someone's Country. This included the Larapinta in the Lake Eyre Basin, also known as the Finke River, which has maintained its present watercourse for at least 100 million years and is likely the oldest in-use riverbed in the world.

Australia's First Peoples thrived in this land when much of what would become North America and northern Europe were under vast ice sheets. Where there were streams and rivers, First Peoples built settlements. They travelled around their Country hunting, fishing, gathering, planting, harvesting. They knew their Country, were an integral part of it, had a storyline for all of it. They respected First Law—customary or traditional law, the lore of the land. The living waters were seen to be interconnected systems, sustained through stories, song, dance, and ceremony. First Peoples flourished because of their connections with, obligations to, and care for Country.

In many written accounts from the nineteenth century, the beauty and pristine nature of Australia's rivers and land were well recognised by those surveying it for settlement. Accounts from Martuwarra, also known as the Fitzroy River catchment, in the Kimberley region of Western Australia, give a sense of what it was like, and the intent of some of those surveying it:

> The water here was beautifully fresh, and the river banks on either side were black with whistler ducks, pelicans, storks, curlews, snipes and other fowl …
>
> Saw good place for a station. I must try and take this country up … I will go right back to Derby if all go well and take this country up as it is too big a thing to lose but at the same time I doubt if anybody will be out here. I had no idea the country was so good. I never saw a better watered creek in my life.[4]

First Peoples' fish traps and weirs for aquaculture are some of the oldest built structures in the world. In the Budj Bim Cultural Landscape, a UNESCO World Heritage site located in the Country of the Gunditjmara people in Victoria, you'll see evidence of thousands of years of continuous aquaculture. At Brewarrina in New South Wales, along the Barwum (also known as the Barwon River), there are Baiame's Ngunnhu, better known as the Brewarrina fish traps, which are many thousands of years old. They testify to how people adapted themselves to the landscape to create accessible and sustainable food sources over millennia.

In Australia's deserts, First Peoples thrived by knowing and moving around Country, by working and sharing together, by understanding nature's water sources, by building and managing rockholes (such as by covering them to reduce evaporation), and by accessing water or moisture from plants and animals. Archaeological evidence shows that First Peoples have been living in Australia's deserts for at least 50 000 years, and, indeed, in every bioregion of Australia for tens of thousands of years.[5] Rock art by First Peoples is among the world's oldest, some of it created many thousands of years ago to tell stories, or depict people and creatures living on Country, including its living waters.[6]

THE GREAT DISPOSSESSION

Many of Australia's rivers were named after European men and settler Australians. The Darling was named after a governor of New South Wales, the Fitzroy in the Kimberley after a captain in the Royal Navy, the Murray after a British secretary of state for the colonies, the Macquarie after another NSW governor, the Daly after a governor of South Australia, and the Brisbane after yet another governor of New South Wales. An exception to these naming conventions is Australia's third-largest river, the Murrumbidgee, which means 'Big Water' in Wiradjuri. But retaining a First Peoples' name did not save it from degradation suffered by its European-named counterparts. Much of the inflow in the Murrumbidgee's upper catchment has been diverted for hydro-electric generation as part of the Snowy scheme, and in 2019 the river stopped flowing just south of Canberra.

The Murray–Darling Basin in south-eastern Australia covers an area as large as Spain, France and Italy combined, and it was where Australia's first 'water wars' took place. Within two decades of the arrival in 1829 of the explorer Charles Sturt, on the Baaka (commonly known as the Lower Darling River), much of the pastoral land along and near its banks had been acquired by squatters.[7] This 'Great Dispossession' was motivated by the desire to acquire highly valuable squatting rights over land—and the water on it—that could support large flocks of sheep, plus the bales of wool were most economically transported to market by river. The Great Dispossession was accelerated by smallpox, which Sturt appeared to have observed among First Peoples along the Baaka; other introduced diseases, like measles, chickenpox and scarlet fever; the destruction of First Peoples' food sources by livestock; the removal of First Peoples' fish traps; and the deliberate poisoning of their food and waterholes to 'move them on'.

And their violent murder. From 1780 to 1930, across all of Australia, there were 421 recorded massacres when at least six people were killed each time. Many of these atrocities occurred at waterholes, streams or rivers. The total estimated number of deaths in these massacres exceeds 11 000.[8] For those carrying out the killings, there were very few consequences. An exception was the Myall Creek Massacre of 1838 in New South Wales, when at least twenty-eight people, including women and children, were murdered. After two trials and much intimidation of the juries, seven defendants were sentenced to hang, while four others walked free. The chief instigator of the massacre, the squatter John Henry Fleming, lived for another fifty years.[9]

By the beginning of the twentieth century, almost all First Peoples' Country suitable for grazing or cultivation had been acquired, without treaty, by one means or another. It wasn't until 1992 that the High Court of Australia recognised the falsehood of terra nullius, and that First Peoples had rights and interests in land and waters according to their own traditional laws and customs on their Country. The federal *Native Title Act 1993* was a direct response to the High Court ruling, and it has provided a mechanism for the determination of native title claims and compensation. The *Native Title Amendment Act 1998*, a federal government response to another High Court decision, ensured that native title rights coexisted on land held by pastoral leaseholders. But in the case of an inconsistency between native title and non-native title, the non-native title rights held sway.[10]

Despite changes brought in with the *Native Title Legislation Amendment Act 2021*, current native title laws in Australia still mean that although First Peoples have some form of ownership over about half of Australia's landmass, these rights give much less control than does freehold title. That is, native title does not necessarily give control over how the land is used or who it is used by, nor does it, by itself, provide water rights to native title holders. Indeed, where there are existing valid leases, permits, licences and the like, including water licences, such instruments can prevail over native title rights and interests.

The deliberate destruction in May 2020 of Juukan Gorge and its artifacts, which were tens of thousands of years old, on the Country of the Puutu Kunti Kurrama and Pinikura peoples, was a wake-up call for Australians about the deficiencies of native title. The wilful obliteration of this sacred site for mining by Rio Tinto, with full knowledge that Juukan Gorge was of the 'highest archaeological significance in Australia', was, in part, a result of deficiencies in native title federally and in Western Australia in particular.[11]

The large native title settlement of 128 000 square kilometres of land in western New South Wales belonging to the Barkandji people in 2015 did not include any water rights to the Baaka, yet the Barkandji call themselves 'People of the Baaka'. And while there have been small returns of water to First Peoples in Victoria since 2020, and $100 million has been promised to buy water rights for First Peoples by the Australian Government, as of late 2024 only 0.2 per cent of surface water rights and 0.02 per cent of groundwater rights in the Murray–Darling Basin were held by First Peoples.

The dispossession continues. In 2024, the NT Supreme Court dismissed a native title challenge by the Mpwerempwer Aboriginal Corporation concerning the granting of a very large groundwater licence, equal to 40 billion litres (enough to fill 16 000 Olympic-size swimming pools) per year, to an agricultural business. Shockingly, the judge in that case found there was no statutory obligation for the water extraction licence to comply with the NT Government's own water plan for the area. This perspective is contrary to principles agreed to by Australian governments, including the NT Government, as part of the 2004 National Water Initiative. As a result, the holder of the pastoral lease, and not the land's native title holders, now has a licence to extract very large amounts of ground water for irrigation in one of the most arid inhabited areas in Australia.

And it gets worse. The NT Government's 2024 draft Mataranka Water Allocation Plan doubles the allowable water diversions. If implemented, it will jeopardise the existence of sacred sites and internationally renowned freshwater springs—for example, Mataranka Springs and Bitter Springs—as well as reduce streamflow along the Roper River.[12] This draft plan was released despite clear evidence of the risks, and strong opposition to greater water diversions by the native title holders of the region, including the Ngalakgan, Alawa, Mangarrayi, Ngandi, Marra, Warndarrang, Nunggubuyu, Ritharrngu-Wagilak and Rembarrnga peoples.

South of the Mataranka water allocation zone is the Georgina Wiso water allocation area. Its water plan was finalised at the end of 2023 and allows for up to 210 billion litres per year of ground water to be extracted, including 148 billion litres for agriculture and 20 billion litres for oil and gas exploration and development. Current water diversions in the Georgina Wiso allocation are 15 billion litres per year, mostly for stock and domestic use. This means that the allowable diversions in the new Georgina Wiso plan are fourteen times larger than was previously the case.

Sadly, the Northern Territory is not the only jurisdiction that is choosing to step back a century in time when it comes to how water is governed. The Productivity Commission Inquiry into National Water Reform in 2024 described how the WA Government, after eighteen years of consultations and drafts, withdrew in December

2023, without any notice, an entire package of much-needed water reform legislation. The consequence is that Western Australia is still in part governed by water planning laws more than a century old, laws drafted when First Peoples were not recognised as citizens and when water was considered simply to be a resource to be extracted.

While not all jurisdictions are as regressive as the Northern Territory—indeed, Victoria is seen as the most progressive—the water reform inquiry of the Productivity Commission observed that interactions by governments with First Peoples are often a 'box ticking' exercise, with little meaningful engagement. What is needed, according to the Committee on Aboriginal and Torres Strait Islander Water Interests, is for First Peoples' key values of water (cultural, spiritual, social, economic and environmental) to be recognised along with First Peoples' cultural rights and interests in water.

Reallocating water to deliver more sustainable outcomes and water justice is to almost everyone's benefit. It is also the best way to manage an increasingly variable and less favourable climate, with predicted greater extremes in rainfall and temperature.

Water reallocation for justice and sustainability is not a utopian vision. The Martuwarra Fitzroy River Council, which represents several First Nations, applies the values, principles and knowledge of Elders to promote such a water future. A systematic and systems view of water, of how it connects to almost everything, is applicable to streams, rivers, wetlands and aquifers. But it requires that we, as a nation, prioritise 'water for all', not just water as an economic resource (water has cultural, spiritual and environmental values) and not just water for humans (water is needed by all fauna and flora). It means we need to change how we manage our water by learning from past mistakes, building on what we know works well, and listening to voices that are too often ignored.

NEGLECT OF RURAL AND REMOTE DRINKING WATER

For many decades, Australians in larger towns and cities have enjoyed, and expected, reliable and safe drinking water delivered to them 24/7, every day of the year. This has happened because of very large investments in water infrastructure, including the protection of catchments that are the source of water for metropolitan centres. This built urban water infrastructure includes well-developed and maintained systems of treatment and disinfection, complemented by frequent monitoring of water quality.

What may come as a surprise to lifelong Melburnians or Sydneysiders is that the care and attention that their water gets is not replicated for the water supplied in many rural and especially remote communities. In 1983, a major study commissioned by the Australian Government highlighted that the water in many small communities was unsatisfactory and needed to be improved 'where necessary and practicable'.[13] Much has changed in the more than forty years since that study; for one thing, Australia's population has increased from 15 million to 27 million people. Yet a 2022 study of publicly available data found that in 2018–19, hundreds of thousands of Australians in regional and remote Australia were still accessing water from their taps, at least for some of the time, that failed the definition of 'good quality water' as per the Australian Drinking Water Guidelines—the national framework introduced in 2011 for describing, managing and monitoring drinking water quality.[14]

This was the case despite Australia signing the 2010 United Nations Resolution 64/292: 'The Human Right to Water and Sanitation'. This resolution recognises 'the right to safe and clean drinking water and sanitation as a human right that is essential for the full enjoyment of life and all human rights'.

The 2022 study also identified major gaps in the monitoring and reporting of drinking water quality across Australia, including hundreds of communities that were self-supplied or serviced by small providers. The largest data gaps by population and geography have occurred where monitoring against the Australian Drinking Water Guidelines took place but, for some reason, was not reported publicly. These data gaps include 1.2 million people served by councils in regional New South Wales, and 141 Aboriginal communities served by the WA Government's Department of Communities.

Such data gaps in terms of drinking water quality are a symptom of broader neglect and a barrier to delivering 'water for all'. These gaps matter because everyone has a right to know what is coming out of their tap from their town water supply. Knowledge is power, and water quality data is empowering. It allows people to engage in informed conversations with service providers, government agencies and political representatives. Such knowledge also provides the opportunity to hold decision-makers accountable for their actions—or inactions. Yet, amazingly for one of the world's wealthiest countries, Australia still does not have a central repository of all drinking water quality data nationwide that guides government decision-making. That is, our national reporting on drinking water services only includes utilities with more than 10 000 connections. And governments can't manage what they don't bother to measure.

So what does this neglect mean for the communities in question? For some, the lack of good-quality water may be due to a high microbial count that makes people sick, and which can be life-threatening for the youngest, the sickest, the most vulnerable. For others, the supplied water may have high levels of nitrates, or salt or trihalomethanes, and in some cases uranium, that exceed the Australian Drinking Water Guidelines. Regularly drinking such water over time can have debilitating and potentially life-threatening health consequences, such as birth defects, and colon and thyroid cancers.

In Wreck Bay, New South Wales, and possibly in dozens of other Australian communities, the surface and ground water are contaminated with elevated levels of perfluoroalkyl and polyfluoroalkyl substances (PFAS) and closely related chemicals known as the so-called 'forever chemicals'.[15] These water contaminants increase the risks of getting cancers and also experiencing heart attacks because, with sufficient exposure, they raise the levels of bad cholesterol in the bloodstream. While there have been efforts to test for these chemicals, and some compensation was paid following a class action suit ($132.7 million was agreed in 2023 to be paid by the Australian Government to about 30 000 claimants[16]), the problem of PFAS contamination remains. Importantly, there has been insufficient effort to make the drinking water safe, or to provide alternative and safer drinking water for affected communities.

A sustainable approach to this longtime neglect of rural and remote drinking water requires that communities be part of the solution.[17] This is essential because there is no one size fits all approach to the delivery of safe water to remote communities. For example, installing a reverse osmosis unit to provide good drinking water will not work effectively, or at all, where water sources have contaminants—for example, by suspended solids or high levels of salts—that damage the membranes of the unit, especially when there is inadequate maintenance. Delivering good-quality drinking water is also about fixing the water-delivery system, which is much more than water storage, pipes and taps. It also includes 'green' infrastructure or the natural environment, which complements 'grey' infrastructure of human-constructed water storage and a network of water disinfection, water distribution and wastewater disposal.

Unfortunately, many of Australia's catchments are not in good health. In 2016, a comprehensive regional scorecard for water quality in Australia's regions found that four regions had inadequate data to be clearly assessed, three could not be assessed as there was no data for assessment, two were in a very poor state (the Murray–Darling Basin and the south-west coast) and two were in a poor state (the north-east coast and the south-east coast)—only two were in a good state (Tasmania and the Tanami–Timor sea coast). An overall assessment of the condition of Australia's water-dependent ecosystems was that they were in a poor state and getting worse.

Australia can learn from overseas communities that have delivered sustainable and improved drinking water quality by incentivising service providers and empowering communities. For example, community members can assist in monitoring water quality while benefiting from access to trained technicians who are paid according to how quickly they fix a broken supply. For communities that have temporary supply failures, safe backup water supplies should be accessible at times of emergency (for example, flooding or extreme drought). Last and not least, ensuring good-quality water means 'caring for country', because surface pollution can negatively affect both surface and ground water, and because substantial upstream overextraction reduces both water quantity and quality.

Given the small size of many Australian remote communities, which often have fewer than 100 people, and the high capital costs involved, some subsidy from larger and better-off communities, especially metropolitan centres, is needed to deliver safe and reliable drinking water for all. A study published in 2024, based on a willingness-to-pay survey of more than 3500 people from across Australia, found that households would, on average, be willing to pay between $324 and $847 per year for ten years to deliver good drinking water to all communities. A similar willingness to pay by all Australians would raise between $8 billion and $33 billion, far exceeding the study's estimated $1.3 billion capital cost needed to deliver good-quality drinking water.[18]

Fortunately, some decision-makers are paying attention to the neglect. For example, in 2023, the Australian Government, through its National Water Grid Fund, committed $150 million over four years to cover the capital costs of water infrastructure to provide safe and reliable water for remote and regional First Peoples' communities; state and territory governments now also need to step up to meet their responsibilities. While the announced federal funding is not sufficient on its own, it is hopefully the first tranche of a greater amount of funding, and other support, towards delivering safe and reliable water for all in Australia.

GREAT VARIABILITY

Australia is the world's driest *inhabitable* continent (Antarctica is simply the driest). Almost one-fifth of Australia is desert, which is a region typically defined by rainfall of less than 250 millimetres per year. Meanwhile, half as much land again in Australia is practically a desert because it gets so little rainfall, combined with high rates of potential evaporation—when it rains, the water evaporates so rapidly that there is little excess surface water to flow into streams or rivers.

At the other extreme, because of Australia's size and latitude, small parts of the continent have lots of rainfall (for example, parts of Cape York and western Tasmania)—on average, more than 3000 millimetres per year. But very few people live in those areas. Despite it being so dry in many places, paradoxically, Australia still ranks in the world's top forty countries ranked by per-capita water availability. This is because of Australia's very low population density.

The extreme spatial variation of water availability in Australia is matched by extreme temporal variation, measured by both rainfall and run-off, the latter defined as the excess surface water that flows over land and eventually reaches streams and rivers. In terms of rainfall, Australia is on average about 50 per cent more variable than the global average, and twice as variable as Europe. Because drier locations are often associated with more variable rainfall, and because Australia has many arid and semi-arid places, the greater temporal variability is even more pronounced in large parts of rural Australia. In terms of run-off, Australia is about 60 per cent more variable than the world average, and some 2.5 times more variable than Europe.[19]

The large spatial and temporal variations in rainfall and run-off have had a big impact on Australia's water story. First, Australia's metropolitan centres are located along a so-called 'Goldilocks' zone where it is neither too wet nor too dry. As early settlers moved out from their original points of arrival, many failed to understand the greater temporal variation (for example, the scale of floods and droughts) in other locations. Even today, some Australians living in metropolitan centres think that what they experience in terms of temporal variability is similar to that in rural Australia. It's not. Rural droughts are often longer, and the floods are larger, than what people experience in Australia's biggest cities.

Failure to recognise the run-off variability and the importance of floodplains and wetlands in inland Australia, especially in the nineteenth century, led to settlements being established in the wrong places (for example, on a floodplain). This pattern unfortunately continues today with some new housing developments, and consequently the homes of hundreds of thousands of Australians, at risk of being flooded. Some towns—such as Lismore, located in a floodplain—will, in the absence of timely interventions, likely experience even greater future variability and flooding due to climate change.

Land-use planning and zoning can help to avoid the worst flooding problems by constraining further development on floodplains in land-scarce cities. An option to better protect people and property already on floodplains is to adapt approaches that have been successful in Australia and elsewhere, such as the Room for the River program of the Netherlands. This program, which began in 2007, seeks to ensure water flows along its natural courses, where floods pose the least risk to lives and infrastructure. Where effective, this approach better protects from flooding the places where people live and work, while creating co-benefits such as the restoration of wetlands.

Second, a key Australian response to high temporal variability has been to build, relative to the country's average rainfall and run-off, much larger water storages than are found in most other parts of the world. For example, to ensure that water storages meet 80 per cent of the expected water demand 95 per cent of the time, Australia needs a water storage capacity about twice as large as the world average, and about six times greater than the European average. Given this, almost all the large dams constructed in Australia have been initiated and paid for by governments, which has increased the importance and impact of Australian governments in terms of water planning.

Third, the large spatial variation in run-off has prompted various inter-basin transfer schemes intended to divert water from relatively wet catchments to relatively dry ones, sometimes hundreds of kilometres distant. These schemes have, in part, been proposed because of the myth that Australia might make its deserts bloom. One of the most famous was the Bradfield Scheme, proposed by an engineer, Sir John Bradfield, in 1938. His original plan was for governments to alter the flow of three Queensland rivers (the Tully, Herbert and Burdekin) that flow into the Coral Sea, and to redirect them westwards to, first, the Flinders River, and then to the Thompson River which, in turn, flows towards Lake Eyre. The key infrastructure of this scheme was to have been a dam 122 metres high and located at Hells Gates on the Burdekin River, which would store water and then have it flow downhill in a 144-kilometre tunnel to the Flinders River. Bradfield claimed his scheme would divert more than 5000 billion litres of water per year into the Thompson River at a cost, in 2020 dollars, of about $3 billion.

The original Bradfield Scheme was recently assessed by the CSIRO, which found it would only have been able to divert about one-third of the volume of water claimed.[20] If the downstream beneficiaries had been charged the full financial costs, not including losses in cultural values and the environmental costs in the source rivers, it would have been unaffordable; that is, the costs would have been multiples more than the actual net returns per litre of water in the destination catchments.

Several versions of the Bradfield Scheme were proposed, including diverting water from the source rivers to the Murray–Darling Basin, where there is a well-established irrigation industry and water infrastructure. A 2020 assessment of this new Bradfield Scheme estimated its capital costs might be as much as $30 billion, with annual pumping and maintenance costs of up to $250 million to divert a maximum of 1270 billion litres of water per year. In CSIRO's best-case scenario, if irrigators had to pay the full capital and operational costs, their additional water costs would have been at least four times greater than the highest value the water could generate in the Murray–Darling Basin.

Fortunately, neither the original nor the new Bradfield schemes were ever built. Nor was another highly ambitious inter-basin water transfer scheme that was still being considered as late as 2012: the Kimberley–Perth Canal/Pipeline. This scheme would have diverted water from Martuwarra in the Kimberley to Perth, via a pipeline some 3700 kilometres long (greater than the distance by road from Adelaide to Darwin). Instead, Perth built two desalination plants (another is planned) to meet its water demands, and at a much lower cost than diverting water from Martuwarra.

Had it been built, the Kimberley–Perth Canal/Pipeline would have topped the Goldfields Water Scheme, which at the time of its construction was the world's longest water pipeline. When it was completed in

1903, it delivered 23 million litres of water per day from Mundaring Dam near Perth to the Kalgoorlie goldfields. While undoubtedly beneficial to residents of the goldfields, an often-ignored impact was the deforestation of thousands of hectares, the trees being cut down to power the steam pumps needed to lift the water to get it to Kalgoorlie.

Fourth, an Australian response to temporal water variability has been to make water more accessible, primarily for agriculture, through the construction of large and publicly funded upstream water storages. The extra water released from these water storages in the summer months provides enormous benefits to those able to access and use this water to grow crops. But almost all the capital costs associated with such large-scale water infrastructure have been paid for by taxpayers, not downstream irrigators. This was a 'sweetheart deal' whereby those who have been privileged with access to additional water have had much of the associated capital costs paid for by state governments.

Over the years, irrigators have not only been given access to water that is unavailable to others, they have mostly been given gratis-free allocations of water in the form of water rights. In New South Wales, for example, past water allocations to licences in irrigation districts included a statutory right of up to three million litres per hectare. Irrigators with a minimum 81 hectares of land suitable for irrigation qualified for the maximum water allocation, which varied from 550 million litres to 743 million litres per year, depending on the irrigation district.[21] These water licences were bundled to the land and were not separately tradeable, but as water rights developed from the 1980s to the 2000s, they became water entitlements, separate to the land and tradeable through a well-developed formal water market.

In today's water markets, a high-reliability water right equal to 550 million litres can be worth more than $1.5 million. Thus, those lucky enough to get free water licences in the past—and smart enough to keep hold of them—have received a very large financial windfall from their state governments. In the Murray–Darling Basin, the total value of water entitlements exceeds $26 billion, with annual turnover of more than $5 billion, most of which is owned by just a few thousand irrigators along with some investors.[22]

The largest beneficiaries of free water rights have, over time, become a 'hydrocracy' akin to the wealthy squattocracy of nineteenth-century Australia, which acquired the right to use, at virtually no cost, vast areas of Crown land by virtue of being the first European settlers to use it. The 21st-century hydrocracy has gone on to wield its influence to impede public-interest water reforms by lobbying public servants, politicians and their advisers to prevent any reallocation of water.

A 2020 investigation by the NSW Independent Commission Against Corruption (ICAC) into alleged inappropriate influences found that water access rules had allowed for 'opportunistic extraction by a small number of large irrigators of unprecedented volumes of water at low flows, which are the flows that are critical to riverine ecosystem health',[23] contrary to the intent of the state's *Water Management Act*. Further, ICAC found that a senior NSW public servant who had provided confidential information to an industry (that is, irrigator) reference group did not act in the public interest and his conduct was improper. As a result of its investigations, ICAC exhorted: 'Public officials cannot be permitted to get too close to one sector in their portfolio and ignore the other stakeholders.'[24]

Some in the hydrocracy are seeking a repeat performance in Australia's north, with their demands for much larger water diversions but without contributing a single cent to the public purse for this valuable privilege. As noted earlier, one business with a pastoral lease on land under native title was recently given the right to eventually extract, at no cost, up to 40 billion litres of ground water per year for thirty years. This water is intended to be used to irrigate up to 3500 hectares of land to grow horticultural crops, such as watermelons. This has gifted the pastoral leaseholder potentially several hundred million dollars worth of scarce but accessible water, courtesy of the NT Government—a handout provided against the express wishes of the traditional owners of the land where the ground water will be extracted.[25]

Fifth, a settler expectation that the temporal rainfall and run-off variability in Australia was comparable to that in Europe, contributed to widespread overstocking and overgrazing in rural areas, especially during the nineteenth and early twentieth centuries. This thinking still exists in the twenty-first century, with some considering droughts to be an exceptional feature of Australia's climate, rather than a standard one for which we must all be prepared and which we can adapt to. When advice was provided about what locations were not suitable for certain types of agriculture, such as South Australia's 1865 Goyder Line that marked the northern rainfall limit beyond which cropping was deemed unsuitable, it was mostly ignored. Some optimistically believed that 'rain follows the plough'. When the inevitable droughts came, many wheat farmers above the Goyder Line were left destitute.

A 'hope for the best' philosophy and/or 'short-termism' was also a contributor to overstocking. This devastated much of western New South Wales and parts of Victoria during extended droughts shortly before Australia became a federation in 1901, with the consequences of such grazing practices leading to profoundly negative changes in the Australian waterscape. Some thought that, because of the widespread devastation of dryland agriculture during extended droughts, governments should step up and fund schemes to drought-proof agriculture through irrigation on riparian land near suitable water sources. What might work for a few, however, does not work for many. Despite the construction of very large water storages in the twentieth century, primarily for irrigation, by the 1970s it was already evident that too much water was being diverted in the Murray–Darling Basin, Australia's 'breadbasket'. Overallocation of water has meant that, during extended droughts, irrigated agriculture is not spared the consequences of too much upstream water extraction and insufficient run-off. And for those downstream irrigators with perennial plantings, such as grapes or fruit trees, if there is no spare water to be had during extreme droughts, the vines or trees on which their livelihoods are based will die.

Finally, there is a 'northern myth', the belief that because there are sizeable areas of tropical Australia that have relatively abundant rainfall—greater than 1000 millimetres per year—there is a large and unmet agricultural potential. Unfortunately, this ignores the very high rates of evaporation in the tropical north, and the highly variable rainfall that renders much of the north unsuitable for irrigation. As far back as the 1960s, it was shown in forensic detail that irrigated agriculture in much of the tropical north is not economically viable *unless* there are very large subsidies from governments, including free water.[26] Yet there continue to be calls for government handouts to build publicly paid-for dams and irrigation infrastructure in the north. The latest came in 2015 with the Australian Government's 'Our North, Our Future White Paper on Developing Northern Australia', which included funding of $200 million for northern water infrastructure, along with $15 million for water resource assessments.

GREAT WATER DIVERSIONS (EARLY YEARS)

Irrigation is about diverting water from storages, streams and rivers, and/or ground water, to supplement water provided from rainfall. Given Australia's highly variable rainfall and run-off, it is not surprising that from early European settlement, at least as far back as the 1830s, there have been schemes to irrigate crops. The benefit to irrigators is that the extra water, over and above any rainfall, helps to overcome a crop water deficit, and thus increases crop yields. Most of the irrigated land in Australia is located near to water sources, such as rivers, and where there is an unmet crop water demand because the potential evapotranspiration exceeds the rainfall in the growing season. Two-thirds of irrigated areas in

Australia can be found in the Murray–Darling Basin, where it has a high value-add, approaching $9 billion in 2017–18.

Irrigation punches above its weight in terms of the large volume of water that it diverts and then consumes. Australia-wide, irrigation accounts for about 70 per cent of the total water diverted, but a greater proportion of the water consumed by people. The reason it accounts for a higher proportion of the water consumed is evapotranspiration, which makes water unavailable in liquid form until it later returns as rainfall as part of the hydrological cycle. By contrast, a much smaller proportion of the water diverted for other purposes, such as for household uses, is consumed than in irrigation. This is because most of the water diverted for households remains in liquid form and ends up as wastewater, meaning it is available for possible reuse with appropriate treatment.

There was widespread political support for subsidised irrigation schemes and dams until at least after World War II. These schemes were known in their earlier years as 'closer settlements' and appeared to offer a win-win. First, it was believed that irrigation would allow new settlers to make the land more productive, thereby increasing prosperity. It was also thought that a greater population in these closer (and more productive) settlements would help overcome the unfounded fear of some that Australia needed to populate or perish. Closer settlements were also associated with the breaking up of the large squatter estates by state governments through government-appointed land boards, which purchased the land at market prices to make it available for smaller, and what was expected to be more productive, farm holdings.

In addition, during World War I, many wanted Australia to become a 'land fit for heroes'. This, in part, was to be accomplished, starting in 1917, by giving veterans the opportunity to access land in what became 'soldier settlements', and that included land suitable for irrigation along the Murray River. More than 10 000 soldiers were subsequently settled after the world wars on closer settlements in both New South Wales and Victoria. In these schemes, soldiers were provided with large up-front loans from state governments, with access to subdivided Crown land.

Large-scale water diversions for irrigation began in Victoria towards the end of the nineteenth century. They were championed by Alfred Deakin, who later became Australia's second prime minister. A critical step, based on Deakin's advice after a study trip he made to the western United States, was Victoria's *Irrigation Act 1886*, followed by similar legislation in New South Wales a decade later. The Victorian Act allowed existing riparian rights to settlers (no rights were allocated to First Peoples) to continue, but required all others (including First Peoples) to have government permission to take and use water beyond 'stock and domestic' uses. The Act allowed for three types of water infrastructure: national works funded and constructed by government, with the costs ultimately to be paid for by irrigation trusts; trust works constructed by irrigation trusts, with loans financed by government based on land values; and private works authorised but not paid for by government. By the mid-1890s there were some twenty irrigation trusts, irrigating some 45 000 hectares of land in Victoria.

Despite its spectacular growth, however, irrigation was not able to pay its way in Victoria. In part, this was aggravated by a severe recession from 1891–95, during which land values plummeted and even banks went bankrupt. At Mildura, on the Murray River, a private irrigation company led by the Chaffey Brothers, who were pioneers in Australian irrigation, filed for bankruptcy in 1895. Around this time, many irrigation trusts were also in financial distress but were saved when the Victorian Government wrote off their loans. But they eventually recovered—by the beginning of World War I, Mildura was a thriving irrigation-based community.

Australia became a federation in 1901, but water remained a state responsibility. To expand irrigation, both Victoria and New South Wales established agencies responsible for the construction of dams and irrigation infrastructure in support of closer settlements. At the federal level in 1914, after years of wrangling, the Murray River Agreement was established, with South Australia insisting on a minimum flow of water at its border to protect navigation on the river. The agreement saw the creation of the River Murray Commission, which committed the Australian Government to both planning and financing irrigation infrastructure, with costs to be shared between New South Wales, South Australia, Victoria and the federal government.

At the top of the River Murray Commission's priorities was the construction of the Hume Dam, completed in 1936. It had a final water-storage capacity of over three billion litres, large enough to irrigate hundreds of thousands of hectares of land along the Murray. Similarly ambitious dam building also happened at the state level. New South Wales completed the construction of the Burrinjuck Dam in 1928 along the Murrumbidgee River, which, with upgrades, now has a capacity of about 1000 billion litres. Victoria completed the Eildon Dam in 1927 (later upgraded) along the Goulburn River, with a capacity of nearly 3300 billion litres, with the primary purpose of irrigating land in northern Victoria. By 1940, the total capacity of Australian large dams, almost all located in New South Wales and Victoria, was almost 9000 billion litres, a 35-fold increase from 1900.[27]

Alongside the boom in storing, diverting and consuming surface water for irrigation was a groundwater boom that had its biggest impact in Queensland. Unlike along the Murray River, this boom was about ground water and pastoralism. The source of this water was the Great Artesian Basin, which stretches over 1.7 million square kilometres to cover large parts of Queensland as well as northern and western New South Wales and South Australia's Eyre Basin. In 1886, a bore in Queensland found water at a depth of more than 500 metres that, through its own pressure, provided up to two million litres per day of ground water. Other artesian bores followed, with more than a thousand operational in Queensland by the start of World War I. This 'opened up' much of Queensland to pastoral farming, and mining, and remains an important water source for town supplies. The Great Artesian Basin's water, however, has high salt levels that make it unsuitable, without treatment, for most irrigated crops.

And as with other great water diversions, there have been downsides. Grazing pressures on previously unstocked land have altered the landscape. As water pressures have declined, some 'mound springs', important cultural sites for First Peoples over millennia, have been destroyed, along with their unique plant and animal life.[28]

GREAT WATER DIVERSIONS (LATER YEARS)

Between 1950 and 1980, Australia continued the rapid expansion of dam building for town and city supplies, hydro-electricity generation and irrigation. As a result, Australia's total water storage capacity now approaches 80 000 billion litres for large water storages (those greater than 10 billion litres). Much of this construction was completed by 1990 and has changed little over the past thirty years.[29]

By contrast, both the number and capacity of farm dams have continued to increase, with a tripling of the total farm storage capacity in the Murray–Darling Basin from the mid-1980s. In 2020, total farm dam capacity in Australia equalled 2700 billion litres.[30] Increased farm dam storage capacity has, in turn, contributed to reduced annual streamflows, especially in drier summer months.

The purpose of building water storages for irrigation is to regulate streamflow so that more water is available in drier months. In the case of large dams, the goal is also to increase water availability in years when there are low inflows. For hydro-electric generation, a key purpose of a dam is to produce electricity and, as much as possible, dispatch it at the highest possible price.

Dams optimised for hydro-electric power generation are typically not managed optimally for irrigation purposes. In multi-purpose dams, there are generally trade-offs between the competing water uses. For example, if the primary purpose is flood control, a dam should be kept sufficiently below its capacity to be able to hold back water when there is a high rainfall event. But if a dam's primary purpose is to supply water for town supplies and/or irrigators, there is an incentive to try to maintain high levels of water storage so that more water is available for dry periods. How these conflicts are managed is built into dam operating rules.

By far the largest water diversion in Australia is the Snowy scheme in the Australian Alps—the source of the Snowy, Murray and Murrumbidgee rivers. The Snowy scheme has been called a nation-building project, and in its heyday it directly and indirectly employed thousands. Construction began in 1949 and the project was fully completed in 1974. The Snowy scheme built sixteen large dams, with a total capacity of 9000 billion litres, and seven hydro-electric power stations. The largest dam is on the Eucumbene River, which flows into the Snowy River. Another dam built as part of the scheme is on the Murrumbidgee River, the Tantangara Dam. It was completed in 1960 and allows for transfers of water, via a tunnel, to Lake Eucumbene for hydro-electric generation.[31] These water transfers to Lake Eucumbene have reduced annual streamflow along the upper reaches of the Murrumbidgee River by about half, and peak flows in the winter months in its upper catchment by about 80 per cent.

The original Snowy scheme, called Snowy 1.0, diverted large volumes of water that previously flowed along the Snowy River to the west, including to the Murray River, for irrigation purposes. Irrigators, however, have not been charged the multibillion-dollar-plus (in current dollars) capital costs that, instead, were to be financed from the sale of electricity. If irrigators had been obliged to fully fund the capital costs, they could not have afforded to irrigate with the additional water that the Snowy scheme provides. Instead, irrigators were expected to pay only the maintenance costs associated with their water deliveries. Nor have irrigators or the electricity generators paid for the environmental and cultural damage resulting from the Snowy River having about 1 per cent of its previous flows because of Snowy 1.0.

As a result of the negative impacts of a greatly reduced streamflow on the Snowy River, there was an intergovernmental commitment to increase its annual streamflow. The stated goal was to increase the Snowy River's average streamflows to 21 per cent of what they were before the Snowy scheme. The *actual* additional flows to the Snowy River, however, have been less—and in some years, much less—than the agreed commitment.[32]

Snowy 2.0, a pumped hydro project announced in 2017, is also claimed to be a nation-building project.

It will eventually divert water along a series of tunnels from the Tantangara Reservoir at 1220 metres elevation to the Talbingo Reservoir at 550 metres. As the water is diverted, electricity will be generated. Water will also be pumped up to the Tantangara Reservoir to act as a source of rapidly dispatchable power, when required. As with many large water-diversion projects, it is well over budget. When Snowy 2.0 was originally announced, it was expected to be delivering power by 2021, at a capital cost of $2 billion. Because of a series of delays, primarily due to tunnelling failures, it is expected to be fully operational in 2028, at a capital cost of $12 billion.

Another so-called nation-building project lies at the other end of Australia: the Ord River Irrigation Project, located in the eastern part of the Kimberley. The first stage was completed in 1972 with the construction of the Ord Dam, which created one of Australia's largest constructed reservoirs at Lake Argyle, with a total capacity of some 5700 billion litres. The Ord project went ahead despite confidential advice to the federal Cabinet in 1966 that it was not worth doing, and detailed public evidence already available in 1965 that it would generate a negative economic return. Nevertheless, the northern myth and nation-building meant that predictions of a negative return were ignored. This warning from 1965 also went unheeded:

> Any crop can be grown in any region at a cost. It is technically feasible to grow pineapples in Antarctica. This book is an attempt to point out to the Australian people that the agricultural techniques which have been developed in tropical Australia are uneconomic and that development there could only proceed at tremendous cost to the nation.[33]

A 1993 cost-benefit analysis of Ord Stage One found that for every public dollar spent, only 17 cents was returned to governments. A calculation of the direct total public and private losses until 1991 (in 1991 dollars) was that taxpayers had spent $613 million in direct project costs to generate $102 million in benefits.[34]

Renewed interest in 'developing' northern Australia in the 2000s led to further investments in Ord Stage Two, in which irrigation infrastructure, at the cost of $334 million (in 2014 dollars) to WA taxpayers, was built to expand the area of land under irrigation by up to 8000 hectares.[35] Much of the irrigated land in the Ord Stage Two project has been planted in pest-resistant cotton. An Ord Stage Three was announced in 2022, with support to be provided by the NT Government but with no accompanying, publicly available cost-benefit analysis. The latest project could irrigate an additional 5000 hectares of land along the Western Australia and Northern Territory border.

Included in the trio of great water diversions of the past half century are the irrigation developments in the Murray–Darling Basin. Many of the largest irrigation schemes, called irrigation districts, are in the southern region of the basin, where both the water storages and water delivery infrastructure have been funded by state governments. These irrigation districts include water infrastructure that allows for bulk diversions of water from the rivers to service many irrigators, through a series of weirs and channels.

In the northern Murray-Darling Basin, irrigators typically divert water directly themselves from 'run of the river' at their own pumping costs, using their own infrastructure. These individual irrigator diversions, however, are facilitated by large upstream dams in the upper catchments of the basin, paid for by the NSW Government. Dams and weirs 'regulate the river' by increasing the reliability and the volume of streamflows in the growing season.

In addition to subsidised water infrastructure, state water-sharing plans determine who gets what water, and when, in a catchment. This has privileged upstream irrigators, often at the expense of downstream communities and town supplies. One such community was Wilcannia, which ran out of town drinking water in 2018–19, during the last drought. This occurred months before it otherwise would have because of large upstream irrigation water diversions.[36] The consequences of such diversions are described by Barkandji ('People of the Baaka') Elder and artist Badger Bates:

> When the European settlers arrived, they wanted to have it all, but Barkandji still survived on the river, and because of the river. Barkandji never left their Country, they are still there, and they love their Country, and it loves them back. But Baaka, and the Barkandji way of life with it, is disappearing because of upstream water extractions.[37]

The principal crop grown in the northern Murray–Darling Basin is cotton. For the entire basin, up to 1232 billion litres was diverted in 2020–21 for growing cotton, equivalent to about 25 per cent of the total water diverted for irrigation that year. In terms of water diversions for the basin, the most important crops in decreasing order of importance, by the volume of water diverted per crop, are cotton, all pastures, fruit and nuts, cereals, rice and grapes. Collectively, the gross value of irrigated agriculture in the Murray–Darling Basin was $8.6 billion in 2017–18.[38]

In New South Wales, bulk irrigation infrastructure was privatised in the 1990s, and former state assets are now owned by unlisted public companies where the shareholders are irrigators and the customers to which these companies supply water. One of the largest is Murrumbidgee Irrigation, which diverts water that is supplied to up to 190 000 hectares of irrigated land and more than 3000 properties. The customers of Murrumbidgee Irrigation pay, on average, a few thousand dollars per year in bulk charges for their water deliveries. As of June 2023, Murrumbidgee Irrigation owned irrigation infrastructure assets worth more than $600 million and could deliver up to 1322 billion litres of water per year to its customers. The market value of the permanent water rights of its customers exceeded $2 billion.[39]

To mitigate the costs to irrigators of planned lower water diversions in the 2012 Murray–Darling Basin Plan, the Australian Government committed to spend $5.8 billion for on-farm and off-farm infrastructure subsidies. This appears to have been a 'Nationals-building' project. In the words of Michael McCormack, deputy prime minister and leader of the National Party in 2019: 'At the end of my political career, I want to be able to point to new dams, bigger weirs, more pipelines, they make so much difference at the local level.'[40]

A stated goal of government-funded water infrastructure and subsidies has been to increase water-use efficiency, specifically the proportion of water that gets consumed growing crops out of the total volume of water diverted by irrigators. Increasing water-use efficiency means that irrigators get more crop yield for each litre of water delivered. It also typically means that a higher proportion of the water diverted for irrigation that was previously not consumed is no longer a 'return flow' through either seepage to ground water and/or surface run-off. These return flows are valuable and support streamflow in rivers and recharge to aquifers. So, while subsidies to increase water-use efficiency are a helping hand to irrigators, they are usually detrimental to downstream communities and reduce streamflows.

Australian governments know that increasing water-use efficiency will typically reduce return flows to rivers. Consequently, when providing their billion-dollar subsidies, they require irrigators to provide, in the form of water rights, up to half of their expected water 'savings'. But whether providing half of the water savings as water rights to increase streamflows is sufficient to deliver a net addition of water to rivers can only be known with proper field-based water measurements, and no such measurements at a farm scale are available. Nor has there been a cost-benefit analysis to ascertain whether the claimed public benefits of subsidising irrigation water infrastructure exceed taxpayers' costs.

A 2019 study of what happened to return flows in the Murray–Darling Basin found there was a much smaller net increase in stream and river flows than is claimed by the Australian Government. Depending on its model parameters, which were based on field water balance data for the Murray–Darling Basin, this study found the *net* addition to streamflow from water-use efficiency subsidies was between 70 billion litres and 210 billion litres per year. By contrast, the Australian Government, without calculating any reductions in return flows, asserts the net gain to be 700 billion litres per year.[41]

Return flows really do matter, and not just for rivers. Australian governments could have acquired the water needed for the environment at a much cheaper price, and with no concerns about the impact on return flows, if they had directly purchased water rights from willing sellers. While such purchases to increase streamflows in the Murray–Darling Basin were part of a structural adjustment paid for by the Australian Government that was announced in 2007, they were halted in 2014. The exception was for closed (by invitation only) tenders—called 'strategic and targeted initiatives'—for specially favoured irrigators. The Australian National Audit Office (ANAO) found that the Australian Government department responsible for these closed tenders 'did not develop a framework designed to maximise value for money … [and] did not negotiate the price for the water entitlements it purchased in all but one instance'.[42]

The ANAO also concluded that probity arrangements for the closed tenders were different (that is, inferior) to open tenders, and that conflicts of interest declarations were not properly documented. Open tenders, where all—not just a select few—irrigators bid to sell their water rights, have been highly effective, and are the least-cost method of acquiring water rights. If the reduction of return flows is also considered, open tenders are up to six times cheaper per litre of water acquired for the environment than through subsidies and water infrastructure projects.

Why would politicians choose to spend additional billions of taxpayer dollars to get a worse environmental outcome? A suitable response may, perhaps, be found in the words of two keen observers of water reform in the Murray–Darling Basin, one of whom was previously a senior official at the Murray–Darling Basin Authority: 'One can only speculate on the degree to which irrigation lobbyists and hydraulic bureaucracies influenced specific policy decisions.'[43]

From 2022 there was a review of the great diversions in the Murray–Darling Basin. This led to the *Water Amendment (Restoring Our Rivers) Act 2023* and a welcome return to the cost-effective open tender process with the planned acquisition in mid-2024 of an additional 70 billion litres of water for the environment from willing sellers. The new Act has also allocated $300 million over four years for communities to prepare for a future with less water.

GREAT DEGRADATION

Since 1788, Australia's waterscape and landscape have been radically transformed. Arguably, the biggest impacts on Australia's rivers have been grazing and the direct removal of vegetation, along with excessive water diversions. These changes to a mainly semi-arid landscape have contributed to Australia having 70 per cent of its river systems as either ephemeral (flowing for only a short period after rainfall) or intermittent (ceasing to flow, with extended dry periods).[44] In some locations, such as the Murray–Darling Basin, land clearing may have increased the sediment load by twenty-fold in some 60 per cent of its river systems. Reduced vegetation has also helped raise the level of saline ground water and, along with irrigation, increased the salinity in many regulated rivers and riparian soils.

The problems of excessive sediment load and pesticides, and an excess of nutrients, are a result of the livestock and cropping practices that exist in many Australian catchments. Some of the more consequential impacts are from the thirty-five catchments that drain into Australia's most iconic natural site, the Great Barrier Reef. According to the Great Barrier Reef Marine Park Authority: 'Poor water quality is a major threat to the Great Barrier Reef, particularly inshore areas. Improving the quality of water entering the Marine Park is critical and urgent.'[45]

In general, the regulation of rivers through the construction of weirs and dams has partially or completely disconnected groundwater and surface interactions. Barriers across rivers have prevented or restricted overland flows to floodplains and wetlands, and damaged and/or destroyed fish, bird and invertebrate habitats.

There is no more obvious sign of the great degradation of Australia's waterways than the fish kills along the Baaka (the Lower Darling River), adjoining the Menindee Lakes, in 2018–19 and again in early 2023. These lakes are naturally occurring and intermittent, normally regularly filling after flooding events in the absence of upstream water diversions. The lakes have existed for thousands of years, but beginning in the 1950s they were 'regulated' with a series of weirs to hold back and store water, and to control the flow of water from and to the Baaka.

The fish-kill events at or near Menindee occurred more than twenty-five years after a cap was placed on surface water diversions in the Murray–Darling Basin; the cap was itself a response to an up to 1200-kilometre-long blue-green algae event along the Baaka in 1991–92 that resulted in a declaration of a state of emergency in New South Wales.[46] The Australian Academy of Science investigated the 2018–19 event, which killed up to two million fish, and found that the direct cause of the deaths was insufficient oxygen in the water. The underlying cause was that 'there is not enough water in the Darling system to avoid catastrophic decline of condition through dry periods'.[47] Not one of the academy's eight recommendations to prevent another fish kill was implemented.

The other mass fish-kill event, in February and March 2023, may have killed as many as 30 million fish. The NSW Government requested that the Office of the Chief Scientist and Engineer investigate this fish kill, which did not occur during a drought. According to the findings:

> Mass fish deaths are symptomatic of degradation of the broader river ecosystem over many years … Explicit environmental protections in existing water management legislation are neither enforced nor reflected in current policy and operations. Water policy and operations focus largely on water volume, not water quality. This failure in policy implementation is the root cause of the decline in the river ecosystem and the consequent fish deaths.[48]

An independent investigation in 2024, in this case into how the NSW Government could improve its water management in the northern Murray–Darling Basin, found:

> There is strong evidence that flows necessary to maintain the health of the rivers and critical ecosystem

> functions are not being met during non-dry times, when there is water available to meet these needs.[49]

These investigations add to a litany of reports and peer-reviewed research that document how many of the rivers and wetlands of the Murray–Darling Basin are in a state of crisis. This neglect, despite the federal *Water Act 2007* and the 2012 Murray–Darling Basin Plan being explicit about the need to ensure an environmentally sustainable level of take (diversion) of water, was explained by the commissioner of the 2019 Murray–Darling Basin Royal Commission to be a consequence of the science that was relied upon, which was

> not best, insofar as it lacked transparency and was unable to be thoroughly tested or replicated. The science was not the best available, insofar as climate change was ignored … Accordingly, the determinations of the Environmentally Sustainable Level of Take and Sustainable Diversion Limit, as at November 2012, were unlawful … the Environmentally Sustainable Level of Take was set at a level which, on the evidence, risks compromising the key environmental priorities prescribed in the Water Act 2007.[50]

In sum, the root cause of the crisis has been a failure in catchment and basin planning due to the undue influence exercised by a privileged few against the public interest. The result is that water diversions in the Murray–Darling Basin remain too great, and insufficient attention has been paid to improving water quality or to the connectivity of flows along river channels, across floodplains, and between surface and ground water. Without water connectivity, key ecological services and functions, such as the transport of nutrients, fish breeding, fish migration and removal of salts, are degraded.

Unfortunately, the degradation of riparian environments is not confined to south-eastern Australia. Successive *State of the Environment* reports, first released in 1995, have consistently identified riparian threats and risks to wetlands that include the overextraction of water, exotic aquatic weeds and pests (for example, the common carp), and pollution.

Based on data from an aerial survey of wetland birds in south-eastern Australia, the average surface area of wetlands over the period 2016–20 was about one-third of what it was over the period 1983–87. Environmental degradation has also occurred even in internationally recognised wetlands such as the Coorong, Lower Lakes and Murray Mouth.[51] The primary cause is that both the timing and volume of streamflows to deliver a range of ecosystem services for the Murray–Darling Basin have not been achieved. According to an evaluation of twenty-three riparian sites and seventy-two requirements across the basin, this lack of water connectivity has resulted in there being only two sites that achieved all their environmental watering requirements, while eleven sites achieved none; and over all sites, 69 per cent of requirements were *not* achieved.[52]

Findings of ongoing river degradation in the Murray–Darling Basin, albeit at the end of the so-called millennium drought (2001–10), were in the last Sustainable Rivers Audit, published in 2012. This audit found that of the twenty-three river valleys it assessed, only two—the Paroo River, with no water diversions for irrigation, and the Warrego River, with virtually no water diversions for irrigation—could receive a pass grade. All the other catchments had either a poor or a very poor health rating.[53] Modelling to assess the impacts of the 2012 Murray–Darling Basin Plan suggests that it may have increased the probability of meeting environmental water requirements from 34 per cent to 46 per cent, but this nevertheless remains a fail grade.

Notwithstanding the ongoing degradation of Australia's rivers and wetlands, there are a few notable conservation successes. The most famous concerned the proposal to build the Franklin Dam as a second stage of a hydro-electric generation project in south-western Tasmania on the Franklin River, adjoining the Gordon River—a project that was cancelled by the Australian

Government in March 1983. The first stage of that project, which was completed on the adjoining Gordon River in 1974, helped to create an environmental protest movement that unsuccessfully sought to protect a unique, natural glacial lake, Lake Pedder, from being submerged by the dam. The later Franklin Dam proposal, in turn, mobilised widespread opposition in a 'No Dams' campaign, and not just in Tasmania. The Franklin Dam featured in the March 1983 federal election, with Labor Opposition leader Bob Hawke declaring that if he were elected, he would stop the dam. He kept his promise.

An equally iconic river, the Martuwarra in Western Australia, has been the target of multiple large-scale water diversions. As noted earlier, there were plans as recently as 2012 to build a pipeline to divert water from it to Perth. Further, as part of water resource assessments commissioned by the Australian Government in 2015, the CSIRO claimed that up to four million hectares of land in the catchment—more than two times larger than the total irrigated area of land in all of Australia in 2020–21—was suitable for irrigation. In wording reminiscent of the northern myth, this CSIRO water resource assessment claimed that up to 170 billion litres of ground water and 1700 billion litres of surface water per year could be diverted for irrigation in the Martuwarra catchment.[54] Fortunately, neither the Kimberley–Perth pipeline nor the possible irrigation diversions proposed by the CSIRO for Martuwarra have proceeded.

Examples do exist of good water conservation in Australia. Melbourne was one of the first cities in the world in the nineteenth century to establish protected catchments that are now located in national parks and state forests where public entry is prohibited. Such entry restrictions reduce the risk of drinking water supplies being contaminated. A similar approach was adopted when establishing the water supply for Canberra, with the purchase of existing landholdings in the early 1900s to protect the water quality in the principal water storage, the Cotter Dam. This has not only conserved riparian ecosystems, it has also reduced the costs of water treatment and the risks of contamination to household water supplies.

The almost decade-long millennium drought that ended in 2010 triggered further water reform in response to widespread environmental degradation that, in part, was a result of diverting too much water for irrigation. These reforms included the 2004 National Water Initiative, an intergovernmental agreement on the principles Australian governments should apply to govern fresh water; the 2007 National Plan on Water Security, which set aside $10 billion (later increased to almost $13 billion) to help ensure, among other goals, an environmentally sustainable level of take in the Murray–Darling Basin through purchases of water rights from willing sellers, and subsidies for water infrastructure and increased water-use efficiency; the *Water Act 2007*, which gave the Australian Government powers in relation to water planning in the Murray–Darling Basin allowing it to 'return to environmentally sustainable levels of extraction for water resources that are overallocated or overused' and 'protect, restore and provide for the ecological values and ecosystem services of the Murray-Darling Basin';[55] and the 2012 Basin Plan that gave effect to the federal *Water Act 2007* and sought to reduce surface water diversions by as much as 2750 billion litres per year.

Collectively, the Murray–Darling Basin reforms have reduced surface water diversions (although the 2012 Basin Plan allowed for increases in groundwater diversions) from what they would otherwise have been, and they have also allowed for more coordinated (although still contentious) water planning in the basin. Water markets have mitigated the costs of reduced water diversions by allowing irrigators with higher market value uses (for example, almonds) to purchase water from irrigators with lower market value uses (for example, pasture). These are useful first steps in water reform, but they are nowhere near good enough. Importantly, a key goal of the *Water Act 2007*, to return to environmentally sustainable levels of extraction for water resources that are overused, has still not been achieved. And climate change will make the water challenges even greater.

FUTURE IMPERFECT

Australia, like every other country in the world, is experiencing climate change. It is also one of the more vulnerable nations because of its susceptibility to global weather oscillations, one of the most important being the El Niño–Southern Oscillation events that contribute to large inter-annual rainfall variability over much of Australia. Climate change in Australia will be experienced mostly through changes to the water cycle, whether it be too much (floods), too little (droughts), or too dirty (sea-level rise) water. These effects will be large. They will also rapidly increase as the world gets closer to a 2°C increase in the average global surface temperature relative to 1850–1900—a likely outcome by 2050 under the current global greenhouse gas emissions trajectory.

In some locations, such as south-western Australia, there has already been a decline in average annual rainfall of about 20 per cent since 1970. A decline in rainfall matters big time for streams and rivers because historical declines in annual rainfall, in some locations in Australia, have resulted in more than a five-fold decrease in streamflow. In other locations, especially in the far north of Australia which already has high rainfall, average rainfall in the wet season is projected to increase with climate change, posing another set of challenges in terms of flooding. A higher average surface temperature, a greater maximum surface temperature, increased annual rainfall variability, and a change in when the rainfall happens during the year, will all adversely impact farming and the environment. Such impacts could reduce the total value of food and fibre production by as much as tens of billions of dollars over what would have been the case in the absence of climate change. In some projections, streamflow in parts of Australia's breadbasket, the Murray–Darling Basin, could decline by as much as 40 per cent over historical levels with a global average surface warming of 2°C.[56] In the absence of a corresponding reduction in water diversions, this will have enormous and negative impacts on already damaged wetlands and river systems.

While Australian farmers have to date successfully adapted to climate change by changing the crops they grow and the dates of planting and harvesting, among other practices, their ability to do so will diminish with much higher maximum temperatures. Given that Australia is a large net food exporter and high per-capita income country, these impacts will not affect food security as much as in poorer, net food importing nations. Nevertheless, the consequences will be large and will likely affect the viability of smaller towns that are heavily reliant on farming enterprises for their employment and income.

Climate change will, without effective adaptation, also make rural water supplies less reliable. Metropolitan cities on the coast of Australia should continue to have reliable water supplies because of desalination plants that were started or built during the millennium drought, and, if necessary, more desalination plants could be constructed, though at a high capital cost. But big-city customers will pay a higher price for their water because desalinated water is more costly to produce than water collected in a dam.

All towns, both large and small, but especially those in more cyclone-prone locations during the summer months, will collectively need to invest billions of dollars in upgrading water infrastructure to cope with more extreme and more frequent high-rainfall events.[57] Management of water storages to meet multiple needs, such as flood control and town water supplies, will also become more difficult as the trade-offs become greater with extended droughts and extreme rainfall events.

Living in a country that already experiences high maximum temperatures, Australian residents are particularly vulnerable to increasing heat extremes. Australia is already experiencing more intense, longer-duration and more frequent high temperatures than it did in 1950, and this trend appears to be accelerating. A key impact of heat extremes is that heat-related fatalities will likely rise, while the quality of life for many, at least in the summer months, will decline. In this climate-change future, Australians are likely to place an even greater value on access to water of sufficient volume and quality for cooling and recreation.

Higher temperatures, and more prolonged droughts, after periods of normal or above-normal rainfall that increase the fuel load, will thereby increase the frequency and intensity of bushfires. In addition to the immediate devastation of bushfires, which includes loss of human and non-human life, vegetation and buildings, there will be additional threats to water quality from sediments (for example, ash, organic material and soil erosion) after fires. Post-fire water contaminants in run-off can change the biota that, in turn, can lower the dissolved oxygen in streams and rivers, and so trigger fish kills. In some locations, as trees eventually grow back and take up more water, there will be less future run-off.

Climate change will exacerbate the trade-offs between competing water uses. This will require much better water planning and a shift away from treating streamflow in rivers as a second-order priority. A suitable adaptation response would be to ensure up-to-date measurements of streamflow and analyses to distinguish between the impact of climate change from increased water diversions, such as from 'floodplain harvesting'. To illustrate, along the Baaka, and as measured at Wilcannia, average streamflow declined from 2314 billion litres per year over the period 1981–2000 to 1087 billion litres per year over the period 2001–20. This has, in turn, contributed to a 75 per cent reduction in the number of waterbirds at the nearby Menindee Lakes, and reduced the resilience of waterbird populations to bounce back following dry periods.[58]

Some have claimed that the streamflow decline along the Baaka over the past twenty years is primarily, or even solely, a result of climate change. If this were correct, it would place the responsibility exclusively on global drivers, for which upstream irrigators cannot be held responsible. There is, however, robust evidence that, at most, 41 per cent of the reduction in streamflow along the Baaka (at Wilcannia) over the past forty years is attributable to climate change, while the rest is attributable to increased water diversions for irrigation. Given that, in dry periods, irrigators in upstream catchments have historically diverted as much as 80 per cent of streamflow, overall water diversions will need to be reduced and river system connectivity increased to avoid an ecological collapse in the northern Murray–Darling Basin with future climate change.

Another aspect of water and climate change in Australia that is not given sufficient planning attention is sea-level rise and storm surges. Depending on the coastal topography and vegetation, a 10-centimetre sea-level rise could increase the frequency of coastal flooding by as much as three times, while a 1-centimetre sea-level rise could result in a shrinkage of coastline by as much as 1 metre. Given that almost 90 per cent of Australians live within 50 kilometres of the coast, the social, environmental and economic impacts of sea-level rise and storm surges will be very large.

In a high emissions scenario, sea-level rise could, on average, exceed 1 metre by 2100. Parts of northern Australia would likely experience an even greater sea-level rise. Some of the largest projected increases in sea level in Australia are for the Kimberley, where coastal freshwater aquifers are at risk of saltwater intrusion—a threat that is magnified by planned increases in groundwater diversions.

Economic damage estimates in 2022 for Victoria only, in relation to sea-level rise and storm surges, project that the costs rise rapidly such that dollar damages will be almost ten times greater in 2100 than in 2040; damage to wetlands will be extensive and could cost as much as $100 billion by 2100; as many as half a million houses in Victoria will be subject to damage, with some 150 000 homes at high risk by 2100; and non-market losses in terms of damaged natural reserves could represent as much as 40 per cent of the total projected economic damages, and could affect an estimated area of 144 000 hectares (about 4 per cent of the current area) and 288 000 hectares of Victorian wetlands.[59]

Notwithstanding the large, and in some cases unavoidable, costs of climate change that will affect Australia in the future, there is much that can be done to reduce damages and losses, and to support resilience to climate risks. This is not about subsidising farmers to stay in places where current agricultural practices are no longer suitable. Instead, it is about preparing for a changing climate, and promoting resilience that allows socio-economic and biophysical systems to bounce back following adverse shocks, but in ways that avoid locking in practices that ultimately make systems less sustainable.

Effective resilience planning includes consideration of each climate risk (for example, flooding) and place (floodplain), and responses to key questions such as adaptation 'for whom' (more vulnerable people and vulnerable communities) and 'through what' (local governments and First Peoples organisations). Such water planning and climate adaptation should be

> comprehensive (across all geographies and hazards), inclusive (engages and listens to all communities), implementable (at multiple scales), integrates (across all sectors and planning), and includes monitoring and evaluation (timely and transparent measurements and

feedback decision loops) to build on successes and to learn from failures.[60]

That is, Australia needs climate adaptation with a resilience framing. This is both necessary and a critically important investment in Australia's future. We can reduce the future costs of climate change with appropriate and adequate top-down funding coupled with comprehensive planning and prioritisation centred around locally based dialogue, knowledge and practices.

VOICES FOR JUSTICE AND SUSTAINABILITY

Retelling Australia's water story is about much more than a history lesson, or a list of what has gone wrong. It is a wake-up call to fix what is broken, to listen to those voices that are seldom heard, and to begin a journey towards a just and sustainable water future for all.[61] The first step on this journey is to put aside what is failing and transform our current water failures into future successes. That is, we need to move from a 'deficit' in terms of how we think and interact with water, to a world that supports an 'abundance' of water for all.

What we need to avoid is summarised by the four Ds:

- Dispossession—the theft of water as part of Country from First Peoples, which began in 1788 and is continuing today with hand-outs of free ground water by the NT Government to agricultural enterprises on native title land, in opposition to the wishes of the traditional owners

- Delusion—the historical failure of many settlers to understand the much greater weather extremes and variability of Australia compared to Europe, and the flaws in the 'northern myth' that there is a huge and unrealised agricultural potential in tropical Australia, and that we can 'drought-proof' this country[62]
- Diversions—taxpayers have dammed, redirected and diverted far more of the available water in Australia compared to water diversions in many other countries, and additional dams and diversions will impose big costs on budgets and the environment alike
- Degradation—the huge cultural and environmental damage done to our waterscapes, especially our rivers and wetlands, by land clearing, water diversions and pollution.

The second step is to think and act differently so that we can move from a grade of D (Deficit) to an A (Abundance). The alternatives are summarised by the four As:

- Aboriginal—a recognition of the primacy of Country, First Law, traditional knowledge and intergenerational care, which have worked successfully for millennia, and that these can guide all of us, not just First Peoples, in how we interact with our living waters

- Accountable—the imperative that decisions about the who, what, where and when of water are made transparently and informed by independently audited water accounts, while ensuring that decision-makers are accountable to act in the public interest: part of a 'democratisation' of water decision-making driven by a balanced bottom-up and top-down process that allows all voices to be listened to
- Adaptable—the recognition that meeting water needs sustainably and equitably requires an iterative and systematic learn-by-doing approach, implemented at the local and catchment scale and which responds to climate change and other threats through robust risk analyses and mitigation actions
- All—access to wholesome water is for everyone, not just the privileged who can afford to pay for the water they want. This means that all Australians must have access to affordable, good-quality drinking water, that wildlife and habitats are respected and appropriately valued, and that delivering water rights to First Peoples becomes a public policy priority.

The third step is to collectively develop shared visions of our water future, and practical actions to get there. These visions must be place-based (for example, at a catchment level) and recognise the values of all, and our relations and obligations to each other. To illustrate, one possible and overarching vision is that

> our rivers, lakes and wetlands have the water needed at the right time to deliver the full set of ecosystem services: [good] water supply for humans; habitat for aquatic and terrestrial animals and plants; water quality and flood regulation; nutrient cycling; recreation; and, importantly, access and use of water by the First Peoples of Australia and Australians living on the land and in cities.[63]

Such a vision is not utopian. And yes, it does unashamedly include justice and sustainability as key objectives. A finer-scale, place-based vision that includes actions to deliver outcomes, in this case for the Murray–Darling Basin, is

> a) restoration of floodplains; b) an increase in environmental water and stream flows at all times needed; c) an audit of the 'shared risks' to water resources in the Basin that include the extraction and storage of water; d) restoration of a Basin-wide environmental and cultural outcomes monitoring program; e) incorporation of appropriate Aboriginal Water Holders to provide certainty of water allocations

> for Aboriginal communities; and f) a requirement that specific actions in anticipation of climate change risks be fully incorporated and accounted for ...[64]

The collective visions of a just and sustainable water future can only be realised by a series of dialogues, and corresponding actions and reflections, which include all voices. Here, dialogue does not mean a 'conversation' or a 'consultation'. Nor does it have to reach a consensus. Rather, dialogue is an ongoing process of listening and responding to voices to collectively decide on agreed-to actions that support a common vision by considering all points of views and the trade-offs to get there.

To illustrate, a dialogue first seeks a common understanding (for example, drivers, pressures, states, impacts and responses) to a common problem (over-extraction of water in a catchment) to then collectively deliberate on the course of actions (catchment-based water planning for all, not just those who are privileged to make a living from water diversions). Dialogue is an ongoing and iterative process that includes action and reflection. Without collective action and reflection, dialogue will not deliver water for all.

Dialogue means that collective water visions, and the actions, reflections and learnings to get there, do not belong to one person or a particular group, and do not privilege one voice over another. It is the antithesis of top-down vision and decision-making. While top-down approaches can result in rapid change and sometimes support the public interest, such as the 2004 National Water Initiative, unless principles and actions are owned by communities at large, they are much less likely to deliver on their goals and thus fall short of what is needed.

How a dialogue would operate needs to be determined by the participants. The dialogue 'ground rules' will differ by place, people and problems, but all ground rules would need to be inclusively formulated, and the process must be ongoing. Dialogue, however, must connect across places and thus should be nested so that local, regional and national problems, and cumulative and shared risks, are fully considered. Dialogue must also bridge different knowledge—Western science, lived experiences, Indigenous science—to triangulate understanding and support robust decision-making. Above all, dialogue must ensure all voices are heard, not just those of the powerful and the influential who typically have access to or can control research funds, the questions that are posed, and the methods used.[65] This means adequate financial and in-kind support for participants so that everyone, including those with limited resources or who are geographically isolated, can be heard. It also means that agreed-to actions at a local level are appropriately funded by governments.

If you think that dialogue is not necessary, that it is a box-ticking exercise or a one-way information-sharing event in Australia's water story, then think again. Our water story, our history of water failures and partial successes, is in large measure a consequence of what has been done by the few haves, those with the power and influence to grab and keep what water they do have. This undermines trust and the needs of the many have-nots who, by contrast, seem to be listened to only when there is a full-blown water crisis, or when an effective response becomes a political necessity.

The need for dialogue is explained by the so-called hydro-illogical cycle.[66] A hydro-illogical cycle manifests itself as business-as-usual until there is an extended drought or other water emergency, at which point Australian governments are forced to act differently and intervene to mitigate 'too much, too little or too dirty' water. But when the rains return, the flood water has receded, or the independent commission has reported back, decision-makers move on to more pressing concerns. This is when those negatively affected by public interest water reform, and who have the most to lose from changes to business-as-usual, push back to impede or even halt agreed-to reforms. Successful pushing back means that externalisation, or the outsourcing of the costs of water diversions to others that has degraded the commons, continues.

The predictive model of the hydro-illogical cycle fits the Australian water story, with multiple examples of reform and push-back cycles. That is, reform occurs in a water emergency, but as the crisis moves from the 'front page' to 'page 8', decision-makers prioritise other issues and fail to follow up on the change needed. Here are some examples of reform and push back, and why dialogue is necessary to deliver a just and sustainable water future.

In the 2010 *Draft Guide to the Murray-Darling Basin Plan*, the proposed Sustainable Diversion Limits were to include a reduction in water diversions to account for climate change. No such reductions appeared in the 2012 Basin Plan. Further, according to the 2019 Murray-Darling Basin Royal Commission, the determination about the environmentally sustainable level of take in the Basin Plan was based on a political compromise and not determined by the 'best available scientific knowledge'. It was also contrary to the key objectives of the *Water Act 2007*.[67]

As another example, no cost-benefit analyses or field-based measurements of return flows were ever undertaken (or at least made available) despite the decision to spend many billions of dollars to upgrade water infrastructure in the Murray–Darling Basin. This is despite decision-makers being informed, in some cases years before, that much of the planned water infrastructure expenditures were neither cost-effective nor likely to achieve their intended increase in streamflow.

Further, during the millennium drought, the Victorian Government built a $625 million north–south pipeline and two pumping stations to divert up to 75 billion litres of water used by irrigators in northern Victoria (who were compensated by $300 million irrigation infrastructure upgrades) as an additional water source for Melbourne. The pipeline was only briefly used in 2010 and has not been used since, ostensibly because it is only needed at times of critical need. Its lack of use, however, has much more to do with political pressure from northern Victorian irrigators and their communities against water diversions to Melbourne.[68]

As a final example, in January 2007, when announcing the National Plan on Water Security, then prime minister John Howard stated: 'As water becomes more scarce and subject to greater demands, it is imperative that we can accurately measure and monitor the resource and its use. This applies equally at the national and Basin scales, as well as for individual farms.'[69] Consequently, the National Plan on Water Security included several hundred million dollars to improve water monitoring and metering. Yet, a decade later, coincident with an ABC investigation in July 2017 into alleged water theft in the northern Murray–Darling

Basin, an independent review panel estimated that between 49 and 75 per cent of surface water diversions in the northern basin were unmetered.

For too long, Australians have been sold a pup, a post-truth story about how the way we manage our water is best practice, and that everyone (or almost everyone) gets a fair go at the water pump.[70] The reality is different. Australia's water story includes great dispossession that remains unresolved today, and great diversions that have disconnected our water systems and contributed to a great degradation of our land and water. And for much too long as a nation, we have neglected residents of small rural and remote communities who live with water that makes them sick. In large measure, this is explained by the haves who have taken and/or misused the water of the commons, sometimes aided and abetted by governments, in turn imposing big costs on the have-nots and compromising our shared water future.

Despite the good intentions of some decision-makers, and the desire by many Australians for a more just and sustainable water future, a hydrocracy of large-scale irrigation enterprises has emerged that is preventing the water reforms that Australia desperately needs.[71] This hydrocracy established itself primarily through the gifting of the capital costs of water infrastructure and free water licences, and then water rights, from the public purse. And what the hydrocracy has achieved in the south of the country, it wishes to repeat in Australia's north.

It's time to push back and deliver the water reforms that Australia must have: a renewed National Water Initiative, regular and independent river and water audits to ensure what is agreed to gets implemented, the restitution of water rights to First Peoples, good drinking water for all, and much more. It's time for all voices to be listened to, not just those of the privileged. It's time for justice to prevail. It's time for a true democratisation of water so that water gets allocated for the benefit of the many, not the few, for today and for the many tomorrows to come.

ACKNOWLEDGEMENTS

At Monash University Publishing, I sincerely thank Greg Bain for the invitation to write this book, and I thank Paul Smitz for his very helpful editing and advice.

I thank all my colleagues, including students, at the Water Justice Hub (https://www.waterjusticehub.org) who have inspired, mentored and supported me on my own water journey. I am also grateful for the support provided to me and the Water Justice Hub by the Hilda John Endowment of the Australian National University, the Australian Research Council Laureate Fellowship Grant FL190100164 'Water Justice: Indigenous Water Valuation and Resilient Decision-Making', and philanthropic funding from Equity Trustees.

As at an acceptance speech, naming people individually can be a problem. What if I were, by mistake, to overlook someone? The thank-you list would include my immediate family—Ariana, Brecon and Carol-Anne, and Henry and Sparky—and all my many (and great) co-authors and friends. However, the genuine appreciation expressed to all who have helped me over the years would be longer than this book! Instead, I limit my named acknowledgements (while nonetheless acknowledging the many who are not named) to those who either read, edited or gave me direct advice about *Retelling Australia's Water Story*. In alphabetical order, they are Matt Colloff, Safa Fanaian, Maurice Nevile, John Williams and Paul Wyrwoll. A very big thank you to all!

I give my respect to all First Peoples' Elders, past and present. I personally acknowledge two Elders: Professor Anne Poelina, Chair of the Martuwarra Fitzroy River Council, and Chair of Indigenous Knowledges, Nulungu Research Institute, University of Notre Dame, and who is a Kimberley, Nyikina Warrwa Indigenous woman; and Honorary Associate Professor Rev. Glenn Loughrey, who is a Wiradjuri man. Anne and Glenn have guided me on my journey to relearn Australia's past to better understand the present.

Let me end at the beginning. Australia's First Peoples have been, and always will be, the traditional custodians of Australia's living waters. We have much to learn from our Elders, and we have much to do to deliver water justice for all!

NOTES

1 RQ Grafton et al., 'The Paradox of Irrigation Efficiency: Higher Efficiency Rarely Reduces Water Consumption', *Science*, vol. 363, no. 6404, 2018.

2 M Kennedy et al., 'Terra Nullius Has Been Overturned. Now We Must Reverse Aqua Nullius and Return Water Rights to First Nations people', *The Conversation*, 30 March 2022, https://theconversation.com/terra-nullius-has-been-overturned-now-we-must-reverse-aqua-nullius-and-return-water-rights-to-first-nations-people-180037 (viewed July 2024).

3 Sydney Water, 'The Tank Stream', fact sheet, n.d., https://www.sydneywater.com.au/content/dam/sydneywater/documents/tank-streamheritage-fact-sheet.pdf (viewed July 2024).

4 Martuwarra RiverOfLife, *Martuwarra Country: A Historical Perspective (1838–present)*, Nulungu reports, 2020, pp. 24, 26, https://researchonline.nd.edu.au/nulungu_reports/2/ (viewed July 2024).

5 J McDonald and P Veth, 'Aboriginal People Lived in Australia's Desert Interior 50 000 Years Ago, Earlier than First Thought', *The Conversation*, 20 September 2018,

https://theconversation.com/aboriginal-people-lived-in-australias-desert-interior-50-000-years-ago-earlier-than-first-thought-102111 (viewed July 2024).

6 P Veth, 'Desert People & Why Heritage Matters', Australian Academy of the Humanities, April 2023, https://humanities.org.au/power-of-the-humanities/desert-people-why-heritage-matters (viewed July 2024).

7 Q Beresford, *Wounded Country: The Murray–Darling Basin—A Contested History*, NewSouth Publishing, Sydney, 2021.

8 University of Newcastle, 'Colonial Frontier Massacres in Australia, 1788–1930', 2022, https://c21ch.newcastle.edu.au/colonialmassacres/introduction.php (viewed July 2024).

9 National Museum of Australia, 'Myall Creek Massacre', 12 April 2023, https://www.nma.gov.au/defining-moments/resources/myall-creek-massacre (viewed July 2024).

10 AIATSIS, *Native Title Information Handbook: National*, 2016, https://aiatsis.gov.au/sites/default/files/research_pub/native_title_information_handbook_2016_national_2.pdf (viewed July 2024).

11 Parliament of Australia, *Never Again: Inquiry into the Destruction of 46 000 Year Old Caves at the Juukan Gorge in the Pilbara Region of Western Australia—Interim Report*, December 2020, https://parlinfo.aph.gov.au/parlInfo/download/committees/reportjnt/024579/toc_pdf/NeverAgain.pdf;fileType=application%2Fpdf (viewed July 2024).

12 M Currell et al., 'Risks in the Current Groundwater Regulation Approach in the Beetaloo Region, Northern Territory, Australia', *Australasian Journal of Water Resources*, vol. 28, no. 1, 2024, pp. 47–63.

13 Steering Committee, in Conjunction with Department of Resources and Energy, *Water 2000: A Perspective on Australia's Water Resources to the Year 2000*, Australian Government Publishing Service, Canberra, 1983, p. x.

14 PR Wyrwoll et al., 'Measuring the Gaps in Drinking Water Quality and Policy across Regional and Remote Australia', *npj Clean Water*, vol. 5, no. 32, 2022.

15 ST Goodwin and K Fuller, 'Commonwealth Reaches $22 Million Settlement with Wreck Bay Aboriginal Community over PFAS Contamination', *ABC News*, 25 May 2023, https://www.abc.net.au/news/2023-05-25/wreck-bay-pfas-compensation/102390538 (viewed July 2024).

16 I Roe, M Taouk and X Gregory, 'Commonwealth Settles $132.7 Million Class Action over PFAS Contamination across Australia', *ABC News*, 15 May 2023, https://www.abc.net.au/news/2023-05-15/pfas-class-action-commonsettlement-reached-with-30-000-claimants/102346274 (viewed July 2024).

17 NL Hall et al., 'Drinking Water Delivery in the Outer Torres Strait Islands: a Case Study Addressing Sustainable Water Issues in Remote Indigenous Communities', *Australasian Journal of Water Resources*, vol. 25, no. 1, 2021, pp. 80–9.

18 A Manero et al., 'Benefits, Costs and Enabling Conditions to Achieve "Water for All" in Rural and Remote Australia', *Nature Water*, vol. 2, 2024, pp. 31–40.

19 DI Smith, *Water in Australia: Resources and Management*, Oxford University Press, Melbourne, 2020.

20 CSIRO, 'An Assessment of the Historic Bradfield Scheme to Divert Water Inland from North Queensland', summary report to the National Water Grid Authority from the

CSIRO Bradfield Scheme Assessment, 2020, file:///C:/Users/pauls/Downloads/20-00418_LW_REPORT_BradfieldSchemeSummary_Historic_WEB_210521_FINAL.pdf (viewed July 2024).

21 W Martin, *Water Policy History on the Murray River*, Southern Riverina Irrigators, Deniliquin, 2005.

22 Productivity Commission, *National Water Reform 2020*, inquiry report no. 96, Canberra, 2021, https://www.pc.gov.au/inquiries/completed/water-reform-2020#report (viewed July 2014).

23 Independent Commission Against Corruption, *Investigation into Complaints of Corruption in the Management of Water in NSW and Systemic Non-Compliance with the* Water Management Act 2000, ICAC, Sydney, 2020, p. 9.

24 Ibid., p. 123.

25 D Connor et al., *Review of the Singleton Horticulture Project's Water Entitlement Provision Costs, Benefits and Employment Impacts*, Centre for Markets, Values and Inclusion, University of South Australia, Adelaide, 2022.

26 B Davidson, *Northern Myth: a Study of the Physical and Economic Limits to Agricultural and Pastoral Development in Tropical Australia*, Melbourne University Press, Carlton, Vic., 1969.

27 DI Smith, *Water in Australia: Resources and Management*, Oxford University Press, Melbourne, 2020.

28 C Harris, 'Five Decades of Watching Mound Springs in South Australia', *Proceedings of the Royal Society of Queensland*, vol. 126, 2020, https://www.royalsocietyqld.org/wp-content/uploads/2020/Proceedings%20126%20Springs/Harris_Web.pdf (viewed July 2024).

29 TA McMahon and C Petheram, 'Australian Dams and Reservoirs within a Global Setting', *Australasian Journal of Water Resources*, vol. 24, no. 1, 2020, pp. 12–35.

30 JL Peña-Arancibia et al., 'Characterising the Regional Growth of on-Farm Storages and Their Implications for Water Resources under a Changing Climate', *Journal of Hydrology*, vol. 625, Part B, October 2023.

31 JJ Pigram, *Australia's Water Resources from Use to Management*, CSIRO Publishing, Collingwood, Vic., 2006.

32 I Bender et al., 'Snowy River Environmental Flows Post-2002: Lessons to Be Learnt', *Marine & Freshwater Research*, vol. 73, no. 4, 2022, pp. 454–68.

33 B Davidson, *Northern Myth: a Study of the Physical and Economic Limits to Agricultural and Pastoral Development in Tropical Australia*, Melbourne University Press, Carlton, Vic., 1969, p. x.

34 Economist at Large, 'Rivers, Rivers, Everywhere: the Ord River Irrigation Area and Economics of Developing Riparian Water Resources', 2014, https://www.ecolarge.com/wp-content/uploads/2014/10/Rivers-Rivers-Everywhere-Developing-Northern-Australia-The-Ord-River-Irrigation-Area-Ecolarge-FINAL.pdf (viewed July 2024).

35 M Grudnoff and R Campbell, *Dam the Expense: The Ord River Irrigation Scheme and the Development of Northern Australia*, discussion paper, The Australia Institute, Canberra, May 2017, https://australiainstitute.org.au/report/dam-the-expense-the-ord-river-irrigation-scheme-and-the-development-of-northern-australia/ (viewed July 2024).

36 M Slattery et al., *Owing down the River: Mortgaging the Future Flows of the Barwon-Darling/Barka River,*

The Australia Institute, Canberra, March 2019, p. 238, https://australiainstitute.org.au/wp-content/uploads/2020/12/P685-Owing-down-the-river-WEB_4.pdf (viewed July 2024).

37 B Bates et al., 'A Tale of Two Rivers—Baaka and Martuwarra, Australia: Shared Voices and Art towards Water Justice', *The Anthropocene Review*, vol. 11, no. 1, 2023.

38 Australian Bureau of Statistics, 'Water Use on Australian Farms', July 2022, https://www.abs.gov.au/statistics/industry/agriculture/water-use-australian-farms/latest-release (viewed July 2024).

39 Murrumbidgee Irrigation, 'Company Overview', October 2023, https://www.mirrigation.com.au/ArticleDocuments/212/MI%20COMPANY%20OVERVIEW_WEB%20READY_FA_Updated%20October%202023.pdf.aspx?embed=Y (viewed July 2024).

40 M Foley, 'McCormack Hails Nats "New Model" for Stunning Election Win', *Mandurah Mail*, 20 May 2019, https://www.mandurahmail.com.au/story/6133792/mccormack-hails-natsnew-model-for-stunning-election-win/?cs=9397 (viewed July 2024).

41 J Williams and RQ Grafton, 'Missing in Action: Possible Effects of Water Recovery on Stream and River Flows in the Murray–Darling Basin, Australia', *Australasian Journal of Water Resources*, vol. 23, no. 2, 2019, pp. 78–87.

42 Australian National Audit Office, *Procurement of Strategic Water Entitlements*, ANAO, Canberra, July 2020, p. 46, https://www.anao.gov.au/work/performance-audit/procurement-strategic-water-entitlements (viewed July 2024).

43 GR Marshall and J Alexandra, 'Institutional Path Dependence and Environmental Water Recovery in Australia's Murray–Darling Basin', *Water Alternatives*, vol. 9, no. 3, 2016, pp. 679–703.
44 M Shanafield et al., 'Australian Non-Perennial Rivers: Global Lessons and Research Opportunities', *Journal of Hydrology*, vol. 634, 2024.
45 Great Barrier Reef Marine Park Authority, 'Land-Based Run-Off', 2022, p. 1, https://www2.gbrmpa.gov.au/land-based-run# (viewed July 2024).
46 C Muir, '"No Triple Bypass, No Miracle Cure, Just a Long Haul Back"' *Inside Story*, 9 October 2014, https://insidestory.org.au/no-triple-bypass-no-miracle-cure-just-a-long-haul-back/ (viewed July 2024).
47 Australian Academy of Science, *Investigation of the Causes of Mass Fish Kills in the Menindee Region NSW over the Summer of 2018–2019*, 2019, p. 2, https://www.science.org.au/supporting-science/science-policy-and-sector-analysis/reports-and-publications/fish-kills-report (viewed July 2024).
48 New South Wales Chief Scientist and Engineer, 'Executive Summary: Independent Review into the Mass Fish Deaths in the Darling-Baaka River at Menindee Findings and Recommendations', 31 August 2023, p. 3, https://www.chiefscientist.nsw.gov.au/__data/assets/pdf_file/0005/580658/Menindee-Fish-Deaths-Report_Findings-and-Recommendations.pdf (viewed July 2024).
49 Connectivity Expert Panel for the NSW Government, *Connectivity Expert Panel Interim Report*, March 2024, p. 5, https://water.dpie.nsw.gov.au/__data/assets/pdf_file/0018/610641/connectivity-expert-panel-interim-report.pdf (viewed July 2024).

50 B Walker, *Murray–Darling Basin Royal Commission Report*, Government of South Australia, 2019, p. 188, https://cdn.environment.sa.gov.au/environment/docs/murray-darling-basin-royal-commission-report.pdf (viewed July 2024).
51 R Kingsford et al., 'A Ramsar Wetland in Crisis: the Coorong, Lower Lakes and Murray Mouth, Australia', *Marine and Freshwater Research*, vol. 62, no. 3, 2011, p. 255.
52 F Sheldon et al., 'Are Environmental Water Requirements Being Met in the Murray–Darling Basin, Australia?', *Marine and Freshwater Research*, vol. 75, May 2024, https://www.publish.csiro.au/MF/pdf/MF23172 (viewed July 2024).
53 Murray–Darling Basin Authority, 'Sustainable Rivers Audit 2: the Ecological Health of Rivers in the Murray–Darling Basin at the End of the Millennium Drought (2008–2010)', fact sheet, 2012, https://www.mdba.gov.au/sites/default/files/publications/SRA2-Interpreting-the-results-factsheet.pdf (viewed July 2024).
54 CSIRO, *Water Resource Assessment for the Fitzroy Catchment: an Overview Report to the Australian Government from the CSIRO Northern Australia Water Resource Assessment*, part of the National Water Infrastructure Development Fund: Water Resource Assessments, CSIRO, 2018, https://www.csiro.au/en/research/natural-environment/water/Water-resource-assessment/NAWRA/Fitzroy-report (viewed July 2024).
55 RQ Grafton, 'Policy Review of Water Reform in the Murray–Darling Basin, Australia: the "Dos" and "Do Nots"', *Australian Journal of Agricultural and Resource Economics*, vol. 63, no. 1, 2019, pp. 116–41.

56 FHS Chiew, 'Climate Change and Water Resources', Munro Oration, Hydrology and Water Resources Symposium Sydney, 13 November 2023, *Australasian Journal of Water Resources*, vol. 28, 2024, pp. 6–17.

57 Climate Council, *Deluge and Drought: Australia's Water Security in a Changing Climate*, 2018, https://research-management.mq.edu.au/ws/portalfiles/portal/133515411/133515243.pdf (viewed July 2024).

58 RQ Grafton et al., 'Resilience to Hydrological Droughts in the Northern Murray–Darling Basin', *Philosophical Transactions of the Royal Society A*, vol. 380, no. 2238, 2022.

59 Victorian Marine and Coastal Council, and Life Saving Victoria, 'A General Summary of the Report: Economic Impacts from Sea Level Rise and Storm Surge in Victoria, Australia over the 21st century (Kompas, T. et al. (2022))', 2023, https://www.marineandcoastalcouncil.vic.gov.au/__data/assets/pdf_file/0036/665649/General-Summary-of-the-Kompas-Report-Economic-Impacts-from-SLR-and-SS-19072023.pdf (viewed July 2024).

60 T Kompas et al., 'Non-Market Value Losses to Coastal Ecosystem Services and Wetlands from Sea-level Rise and Storm Surge, 2050 to 2100: the Kimberley Region, Western Australia', *Ocean and Coastal Management* vol. 255, 2024, p. 7.

61 L Hartwig et al., 'Water Colonialism and Indigenous Water Justice in South-Eastern Australia', *International Journal of Water Resources Development*, vol. 38, no. 1, 2021, pp. 30–63.

62 J Williams, 'Can We Myth-Proof Australia?', *Australian Science*, vol. 24, no. 1, Jan/Feb 2003, pp. 40–2.

63 Crawford School Policy Brief, 'Water Reform for All: a National Response to a Water Emergency', Australian National University, Canberra, 2020, p. 3.
64 Ibid., p. 12.
65 SA Williams et al., 'Diversity, Equity, Inclusion, and Justice in Water Dialogues: A Review and Conceptualization', *Journal of Contemporary Water Research & Education*, vol. 177, 2023, pp. 113–39.
66 DA Wilhite, 'Breaking the Hydro-Illogical Cycle: Changing the Paradigm for Drought Management', *EARTH Magazine*, vol. 57, no. 7, 2012, https://core.ac.uk/download/pdf/33146422.pdf (viewed July 2024).
67 Government of South Australia, *Response to the Murray–Darling Basin Royal Commission Report*, September 2023, https://cdn.environment.sa.gov.au/environment/docs/Murray-Darling-Basin-Royal-Commission-response-report-for-online-viewing2023.pdf (viewed July 2024).
68 L Crase et al., 'The Closure of Melbourne's North–South Pipeline: a Case of Hydraulic Autarky', *Economic Papers*, vol. 33, no. 2, 2014, pp. 115–22.
69 RQ Grafton, 'Policy Review of Water Reform in the Murray–Darling Basin, Australia: the "Dos" and "Do Nots"', *Australian Journal of Agricultural and Resource Economics*, vol. 63, no. 1, 2019, p. 133.
70 RQ Grafton et al., 'Confronting a Post-Truth Water World in the Murray–Darling Basin, Australia', *Water Alternatives*, vol. 13, no. 1, 2020, pp. 1–28.
71 RQ Grafton and J Williams, 'Rent-Seeking Behaviour and Regulatory Capture in the Murray–Darling Basin, Australia', *International Journal of Water Resources Development*, vol. 36, nos 2–3, 2020, pp. 484–504.

IN THE NATIONAL INTEREST

Other books on the issues that matter:

David Anderson *Now More than Ever: Australia's ABC*
Kevin Bell *Housing: The Great Australian Right*
Bill Bowtell *Unmasked: The Politics of Pandemics*
Michael Bradley *System Failure: The Silencing of Rape Survivors*
Melissa Castan & Lynette Russell *Time to Listen: An Indigenous Voice to Parliament*
Inala Cooper *Marrul: Aboriginal Identity & the Fight for Rights*
Kim Cornish *The Post-Pandemic Child*
Samantha Crompvoets *Blood Lust, Trust & Blame*
Satyajit Das *Fortune's Fool: Australia's Choices*
Richard Denniss *Big: The Role of the State in the Modern Economy*
Rachel Doyle *Power & Consent*
Jo Dyer *Burning Down the House: Reconstructing Modern Politics*
Wayne Errington & Peter van Onselen *Who Dares Loses: Pariah Policies*
Gareth Evans *Good International Citizenship: The Case for Decency*
Paul Farrell *Gladys: A Leader's Undoing*
Kate Fitz-Gibbon *Our National Crisis: Violence against Women & Children*
Kate Fitz-Gibbon *Our National Shame: Violence against Women*
Paul Fletcher *Governing in the Internet Age*
Carrillo Gantner *Dismal Diplomacy, Disposable Sovereignty: Our Problem with China & America*
Jill Hennessy *Respect*

(continued from previous page)

Lucinda Holdforth *21st-Century Virtues: How They Are Failing Our Democracy*
Simon Holmes à Court *The Big Teal*
Andrew Jaspan & Lachlan Guselli *The Consultancy Conundrum: The Hollowing Out of the Public Sector*
Andrew Leigh *Fair Game: Lessons from Sport for a Fairer Society & a Stronger Economy*
Ian Lowe *Australia on the Brink: Avoiding Environmental Ruin*
John Lyons *Dateline Jerusalem: Journalism's Toughest Assignment*
Richard Marles *Tides that Bind: Australia in the Pacific*
Fiona McLeod *Easy Lies & Influence*
Michael Mintrom *Advancing Human Rights*
Louise Newman *Rape Culture*
Martin Parkinson *A Decade of Drift*
Jennifer Rayner *Climate Clangers: The Bad Ideas Blocking Real Action*
Isabelle Reinecke *Courting Power: Law, Democracy & the Public Interest in Australia*
Abul Rizvi *Population Shock*
Kevin Rudd *The Case for Courage*
Don Russell *Leadership*
Scott Ryan *Challenging Politics*
Ronli Sifris *Towards Reproductive Justice*
Kate Thwaites & Jenny Macklin *Enough Is Enough*
Simon Wilkie *The Digital Revolution: A Survival Guide*
Carla Wilshire *Time to Reboot: Feminism in the Algorithm Age*
Campbell Wilson *Living with AI*